Spring pool : 13214
J 574.5 DOW

Downer, Ann,
Plaistow Public Library

PLAISTOW PUBLIC LIBRARY
14 ELM ST. P.O. BOX 136
PLAISTOW, N.H. 03865

```
J        Downer, Ann
574.5       Spring pool
DOW
```

JUN 15 '93

MAY 16 '94

Spring Pool

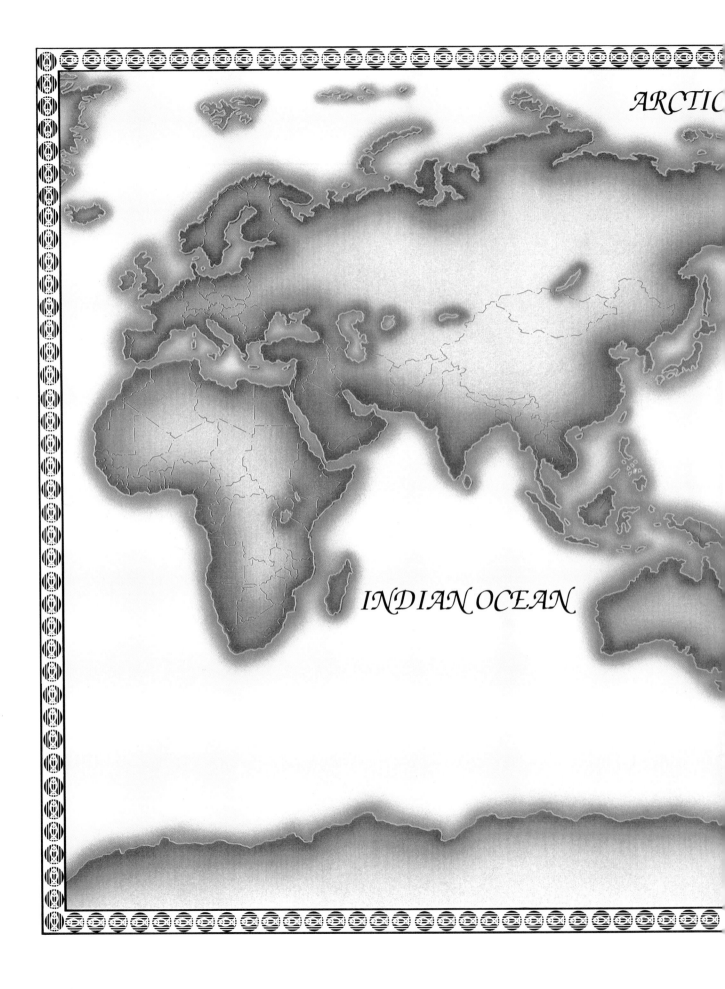

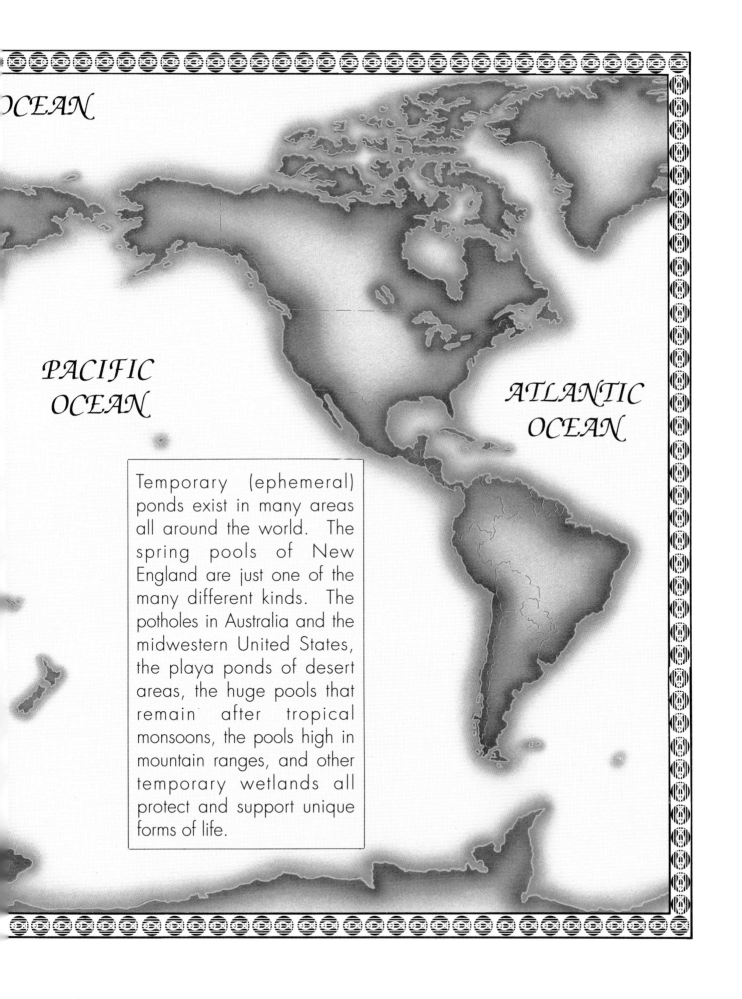

PACIFIC OCEAN

ATLANTIC OCEAN

Temporary (ephemeral) ponds exist in many areas all around the world. The spring pools of New England are just one of the many different kinds. The potholes in Australia and the midwestern United States, the playa ponds of desert areas, the huge pools that remain after tropical monsoons, the pools high in mountain ranges, and other temporary wetlands all protect and support unique forms of life.

Spring Pool

A Guide to the Ecology of Temporary Ponds

BY ANN DOWNER

Endangered Habitats
A New England Aquarium Book

FRANKLIN WATTS
NEW YORK ○ CHICAGO ○ LONDON ○ TORONTO ○ SYDNEY

*This book is dedicated to the children of
Amherst, Massachusetts.*

ACKNOWLEDGMENTS

Special thanks to the following for their kind assistance:
Henry Wilbur, Duke University
Brian Windmiller, Tufts University
Henry Woolsey, Natural Heritage and Endangered Species Program
Betsy Colburn, Marilyn Flor, Scott Jackson, Elissa Landre, and Tom Tyning,
of the Massachusetts Audubon Society.

Author's Note: The pool used as an example in this book is in eastern Massachusetts. The changes and events described here may occur a little earlier or later in the year at pools in other locations.

World map endsheets by Joe LeMonnier

Photographs copyright ©: Tom Tyning: pp. frontis, 33, 37; Scott D. Jackson: pp. 4, 5, 7 left, 10 top right, 10 third from top, 12 bottom, 14 bottom, 17 bottom, 18 center, 23 bottom, 26, 44 bottom right, 46 top right & both bottom, 47 both left, 49 top right; Ted Levin: pp. 6 top, 13 both right, 15, 16, 17 top, 34, 48 top right; Kenneth Mallory: pp. 6 bottom, 12 top right, 13 left, 14 both top, 21 top, 28 top, 29, 30 top, 32, 36, 39, 43 top right; Massachusetts National Heritage Endangered Species Program: pp. 7 right, 8 center right, 44 top right, 46 top left, 48 bottom right (all Steven Roble), 31, 35, 38, 41 (Mark Blazis), 44 left (Scott Jackson), 45 right, 48 top left (Bill Byrne); Massachusetts Audubon Society: pp. 8 top left (Jose Garcia), 11 (Marilyn Flor), 12 top left, 25 top (Gail King), 25 center (Betsy Coburn), 42 top (Wilfred Kimber); Ursus Inc.: pp. 8 top right, 20 both center & bottom, 22, 43 top left (all Ernie Cooper), 8 center left (Finn Larsen), 10 left (P. Hulbert); Photo Researchers Inc.: pp. 8 bottom (John Bova), 12 center (J.L. Lepore), 19 (Nuridsany/Perennou), 20 top left, 43 bottom left, 45 bottom left (all Stephen Dalton), 21 bottom (M.I. Walker), 23 top (Biophoto Associates), 24 (Noble Proctor), 25 bottom (Patrick Grace), 27 (John Walsh/SPL), 28 bottom (J.C. Thompson/NAS), 42 bottom (Herbert Schwind), 45 top left (Sinclair Stammers/SPL), 47 bottom right (Leonard Lee Rue III/NAS), 49 bottom left (J. Serrao); Photo/Nats Inc.: pp. 9, 10 bottom right, 18 bottom, 49 top left (all Stephen Maka), 10 second from top, 18 top, 50 bottom left (all David Stone), 30 bottom (Jo-Ann Ordano), 47 top right (Herbert Parsons), 48 bottom left (Priscilla Connell), 49 bottom right (Valorie Hodgson), 50 top (John O'Connor), 50 bottom right (Don Johnston), 51 left (Gay Bumgarner); Greg Neise: p. 20 top right; Johann Schumacher Design: p. 40; Ian J. Adams: p. 43 bottom right; Scott W. Hanrahan: p. 51 right.

Library of Congress Cataloging-in-Publication Data

Downer, Ann, 1960-
Spring pool: a guide to the ecology of temporary ponds/by Ann Downer.
p. cm.
"A New England Aquarium book, endangered habitats." Includes bibliographical references and index. Summary: A close-up look at ponds formed by melting snow and spring rains and at the varied life that exists in them. ISBN 0-531-11150-4 (library).—ISBN 0-531-15251-0 (trade) 1. Vernal pool ecology—Juvenile literature. 2. Pond fauna—Juvenile literature. 3. Vernal pool ecology—Massachusetts—Juvenile literature. 4. Pond fauna—Massachusetts—Juvenile literature. [1. Pond ecology. 2. Pond animals.] I. Title. QH541.5.P63D69   1992
574.5'26322—dc20   92-19269 CIP AC

Copyright © 1992 by the New England Aquarium
All rights reserved
Printed in the United States of America
6 5 4 3 2 1

Contents

A Spring Pool Comes to Life	5
Epilogue: Slow—Salamander Crossing!	32
Keeping a Spring Pool Diary	34
Field Guide to Spring Pool Ecology	42
Glossary	52
Bibliography	53
Read More About It	54
Index	55

Most spring pools dry up in the heat of summer, then may fill again in the fall and remain as ice-capped pools until the following spring. Vernal pools are usually no more than 3 feet deep and 150 feet across.

A Spring Pool Comes to Life

Life is harsh for animals that winter over in a spring pool. Small animals such as daphnia and fairy shrimp go into a dormant stage until spring rains arrive.

It is late March, and a spring storm passes over a forest in New England. Cold rain glistens on the leaves littering the forest floor and hangs in heavy silver drops from the twigs and branches of trees just beginning to bud. This rain heralds the end of a long winter. The ground is beginning to thaw, and before long the woods will be green once more. In a small clearing, melting snow has filled a leaf-lined hollow, forming a shallow pool. Mist rises from its surface, and over the sound of rain on dead leaves, wood frogs are quacking like ducks. A spring pool is coming to life.

Because most are formed in the spring—by melting snow and spring rains—these ponds are called vernal, or spring, pools. Through the summer, spring pools teem with many forms of **aquatic** life, and by its end most shrink to shallow mud holes, or depressions in grassy meadows. When spring rains return, the pools are refilled and the cycle begins again.

Melting snows fill the depressions, or kettle holes, in woods, meadows, sand flats, and plains. Rains and snowmelt revive dormant life in the pools.

As winter changes into spring, the leaf litter lining the spring pool comes alive with insects and a host of other animals.

FOR SOME, A SAFE PLACE TO GROW

By the end of summer, spring pools are very shallow. Much of the oxygen in their water is lost, and sometimes they dry up completely. Because of this, fish cannot live in spring pools. With no hungry fish to attack eggs and **larvae**, spring pools are ideal breeding grounds for many **insects** and **amphibians**, especially those that take a long time to become adults.

A chick or a puppy always resembles the hen or golden retriever it will grow up to be, but the eggs of many insects and amphibians hatch into a special in-between form that looks very different from the adult. In some animals this form is called a larva; in others it is called a **nymph**. Caterpillars and tadpoles are familiar examples of larvae. Dragonflies are an example of an animal with a nymph stage. Larvae are often too small or slow to escape predators (often other larvae) that try to eat them. Most insects and amphibians make up for this by laying eggs by the hundreds or even thousands so that at least some may survive. The larva or nymph will spend weeks, months, or even years just eating, before transforming into an adult. In the meantime, the spring pool provides shelter and food, away from the fish and other predators found in larger ponds. **Species** that can breed in vernal pools and nowhere else are called **obligate**, or dependent, species.

Among signs of spring in the vernal pool are masses of eggs deposited by spotted salamanders. Clumps of eggs are suspended from submerged twigs or stems close to the surface so that the developing young are supplied with oxygen and are warmed by the sun.

Wood frog eggs are another early sign of life in a spring pool. You may not see the eggs, but you'll know frogs are near by their raucous calls, sounding like a family of quarreling ducks.

Once hatched from protective egg casings, salamander larvae are aquatic for two or three months. They breathe with a set of branching gills, and feed on tiny animals at the pool bottom.

Dragonfly nymphs are among the many animals that prey on salamander larvae. This nymph will eventually change from an underwater larval form to a terrestrial, air-breathing adult.

For three or four months, pool activity is feverish since the water may disappear when summer heat arrives. Here the larva of a caddisfly dines on a tadpole.

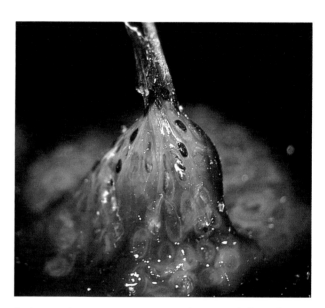

As late spring temperatures rise, the animals face the threat of receding waters. The jelly mass that surrounds and insulates this egg mass helps the embryos inside to survive out of water for up to a week.

Caterpillars and tadpoles are examples of larvae, an early stage in some animals' life cycles. Before the caterpillar becomes a butterfly, it may use vegetation around the spring pool for food and shelter.

FOR OTHERS, A REST STOP

Although they don't breed in the spring pools, raccoons, herons, and other animals use them as rest stops, for food and shelter along the forest "highway" between larger ponds, streams, and lakes. These animals and others use spring pools but don't depend on them. They are nonobligate, or **facultative**, species.

As long as there is enough rain or snow, the pools reappear spring after spring. That's nature's scheme. Unfortunately, humans have upset this fragile balance. As more and more wild places vanish to make room for highways, shopping centers, and office parks, spring pools vanish with them. Later we'll see how spring pools are being threatened. But first let's look at some of the animals that live in spring pools and see how they have adapted to life in this special environment.

A view from an airplane would show that spring pools form links in a water chain connecting wetlands across the country. Animals migrating in search of food, territory, or mates use these oases for shelter and food. One common visitor is the great blue heron.

Other frequent visitors to spring pools are the raccoon (above) and (from the top of the page) Eastern spadefoot, snail, American toad, and red spotted newt.

SALAMANDERS ON THE MARCH

A good way to find a vernal pool is to follow the salamanders. On one of the first wet nights in the spring, when the frogs are singing, shine a flashlight on the leaves that litter the forest floor. With luck, you will see a spotted salamander making its way to the breeding pool.

Students brave the cold and damp, hoping to discover spotted salamanders making their annual trek to spring pools to mate. Salamander enthusiasts are careful not to disturb the animals or their eggs so that the salamanders can return the next year.

The first soaking rains of spring rouse spotted salamanders in the underground burrows where they spend the winter. Obeying some signal—temperature, moisture, or an internal clock—they emerge and begin their journey which can be just a few dozen yards, or as long as half a mile. The salamanders may cross fields and highways to reach their breeding pool. The pool is often the one where the salamander was hatched or, rarely, one selected by the salamander as an adult. Once the pool is chosen, the salamander returns to it every spring for the rest of its life—for some, as long as thirty years.

Males usually arrive at the pool first and are soon joined by the females. The salamanders gather at the spring pool for several days and nights, long enough to lay and fertilize eggs. In a mass mating ritual called **congressing**, salamanders wiggle and slide over, under, and past one another, searching for mates. Once the males find mates, each deposits a pile of jelly called a **spermatophore** on a submerged twig or leaf.

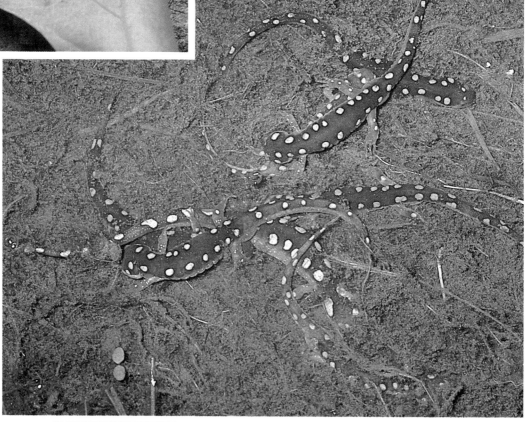

During the first warm, rainy, spring night, spotted salamanders emerge from winter burrows in the forest floor and make their way to the spring pool.

Dozens of spotted salamanders gather in the spring pool in the mating ritual called congressing.

The white dots on the leaves (seen in a close-up view) are sperm packets, which the females grasp with their cloacae to fertilize their eggs.

Each spermatophore encloses sperm, the sex cells that contain the male salamander's genes. Genes are special molecules that carry instructions in chemical code for making new salamanders. When the male's genes combine with the female's genes inside each egg, a new salamander will begin to form.

But first the sperm and egg have to get together. Male and female salamanders have special openings in their undersides called **cloacae**. The female uses folds of skin on either side of the opening to grip the spermatophore and place it inside the cloaca, where the eggs will be fertilized as they are laid.

The female lays up to 250 eggs (depending on the species) that are held together in a softball-sized clump called a **matrix**. The matrix can be clear, milky, or green and it has the feel of Jell-O. It cushions the eggs and holds them together so that they don't drift away. The female suspends the egg mass from a submerged twig.

Once the eggs are laid, the salamanders wait for the next wet night to make their way back to the safety of the leaf litter on the forest floor. Not all of the salamanders will survive the return trip: some will fall prey to birds, snakes, and other predators; others will be run over by cars.

Above: After several days, the female salamander attaches the fertilized egg mass to a submerged twig.

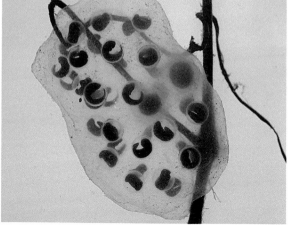

Top, right: Depending on the species, female salamanders that use vernal ponds produce from 60 to 500 eggs in a breeding season. Here, embryos are developing in egg capsules.

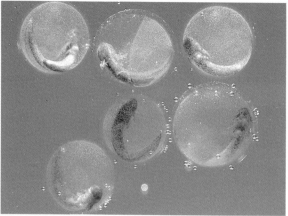

Below: Often the egg mass of the spotted salamander is dyed green. This is caused by the presence of a species of algae.

The following dawn, the spring pool appears deserted. But underwater a remarkable transformation has begun. As the weeks pass, the embryos inside the eggs take shape. Soon tails, eyes, and gills are clearly visible. Six to eight weeks later, the salamander larvae wriggle free from their egg casings. Their lungs are still forming; for now they will use gills to breathe oxygen dissolved in the water of the pool. By June, if they escape the jaws of diving beetle larvae and other predators, the salamander larvae will be well on their way to becoming adults.

A salamander larva hatches and then, during its first weeks of life, preys on tiny aquatic animals. The larva uses feathery gills to breathe under water. The gills will shrink and eventually the larva will become a terrestrial adult, using lungs to breathe.

Predators such as the garter snake (eating a wood frog) and the water snake lie in wait for salamanders journeying from the mating pool to the protection of leaf litter on the forest floor.

A CHORUS OF FROGS

Salamanders are almost mute (they do occasionally chirp), but frogs and toads make up for their quiet relatives. A spring pool at dusk during the frog breeding months of March and April is a raucous place. Sometimes the noise seems deafening, with whistles, clucks, and trills coming from a chorus of many kinds of frogs and toads.

Frogs have four distinct calls: mating, territorial, release, and alarm. Males use mating calls to boast to females of their species that they are the best mates at the pond. A territorial call is directed at rival males. It means "This is my territory. Keep out or else!" A release call is used by both males and females when grabbed by a male. It can mean either "Let me go. I'm a male." or "Sorry, I've already laid my eggs for this season." Frogs give an alarm call when they are threatened. You will hear this call if you accidentally startle a frog. It will give a short squeal before diving into the safety of the pool.

SPRING PEEPER

GREEN FROG

GRAY TREE FROG

All these frogs—seen with vocal sacs swelling as they call or sing—are occasional spring-pool visitors. The males sing to attract females and scare away rival males. Both sexes utter alarm and release calls, roughly translated as "leave me alone" or "get off my back."

AMERICAN TOAD

When you think of a frog, you probably picture it sitting on a lilypad or basking on a half-submerged log—somewhere near water. That's only partly true. Surprising as it may seem, frogs and other amphibians spend most of their lives on land, traveling great distances in search of food. Unlike a snake's scales or a turtle's shell, an amphibian's skin is thin and moist. Frogs have two solutions to the problem of drying out on land: patches of skin inside the frog's thighs that soak up water like a sponge, and a bladder that stores water.

Wood frogs, unlike most frog species, depend on spring pools for a place to lay their eggs. They sometimes arrive at the pool even before the salamanders. In an embrace called amplexus (from the Latin "to twine around"), males fertilize the eggs as the female lays them in a loose mass of 1,000 or more.

Spring peeper egg cases (right) and gray tree frog eggs (below) are often found in the same pond. Neither species depends on the spring pool, but they may be seen there frequently.

The American toad is the familiar "hoptoad" or common toad. In this courtship embrace, sperm and eggs unite. Females then deposit 12,000 eggs in long, black, pearl-like strands around pool vegetation.

LIFE ON THE SURFACE

Whirligig beetles are most visible at the surface where they spin and bounce off each other. They occasionally dive, using air trapped in a bubble beneath their abdomens. Whirligig beetles have four eyes; one pair sees underwater, the other is for surface vision.

From a distance, a spring pool seems tranquil, but a closer look reveals a surface alive with activity. Shiny black whirligig beetles spin and collide on the surface, reminding pond watchers of bumper cars at a fairground. Small, wingless insects called springtails use hairlike triggers under their tails to push off from the water and jump into the air. A diving beetle surfaces to renew the supply of air stored in the space beneath its shell. A pond skater races across the water, its long legs leaving dimples on the surface.

All the frenzied activity has a purpose, and usually that purpose is to find and catch prey. Whirligig beetles use two pairs of eyes for spotting prey on or just below the surface. The backswimmer uses its rear legs to move itself upside down through the water. This swimming style leaves its front legs free to grab tadpoles and salamander larvae. A pond skater walks on water, feeling for vibrations on the pool's surface film the way the spider waits for a tug on the strands of its web. At the first tremor, it dashes over the surface and seizes its prey in its claws. These are set back on its two front feet so that they won't break the surface film.

Diving beetles are voracious predators, both as larvae and adults. This larva is devouring the remains of a tadpole.

Giant water bugs can be 3 inches long and will tackle frogs, tadpoles, and other prey many times their size. Like other water insects, they carry oxygen in an air bubble and breathe while under water through the tip of the abdomen.

Water boatmen live in habitats as different as spring pools, ponds, puddles, and even birdbaths, where they eat algae. Adults can fly, but are most often seen "rowing" through the water with their oarlike middle and hind legs.

Backswimmers spend their lives belly-side-up, just below the surface. Waiting to prey on insects and tadpoles, backswimmers cling to underwater vegetation breathing oxygen they have trapped in air bubbles.

Spring pools are one of many habitats mosquitoes use for breeding. Their larvae (seen breathing at the pool surface) can be found in any still body of water. In spring pools, they are an important food source for aquatic insects.

IN THE WATERS OF THE POOL

A small sample of spring pool water may reveal thousands of tiny animals.

Below the surface, the water teems with salamander and insect larvae, tadpoles, and hundreds of other animals, many transparent and small enough to swim in a teaspoon. Some of these are called **zooplankton**, from Greek words for "tiny drifting organisms" and 200,000 could fit in a glass of water.

Zooplankton are very important to all the life in the spring pool. They are food for the insects that, in turn, feed growing salamanders and other **vertebrate** predators. There are hundreds of kinds of zooplankton, each marvelously adapted for life in the spring pool. We will look at two—the water flea and the copepod.

One of the most common animals in pond water is the water flea, sometimes called daphnia. It's not a flea at all but a member of the group of aquatic animals that includes shrimp. Water fleas use their legs to strain bacteria and algae from

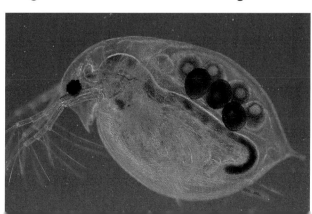

Water fleas are an important strand in the food web of the spring pool. The pinhead-sized daphnia (seen magnified here, carrying eggs) live only about 100 days. But there are so many, and they reproduce so quickly, that they are a major food source for larvae of other species.

the water and pass them to their mouths. Females carry eggs in a pouch beneath their shells. After the young hatch, they remain in the pouch for several days, feeding on a milky dew produced by the female.

Copepods are part of a large family of crustaceans. One kind of copepod (pronounced COH-puh-pod) gets its name from a Greek legend. It is named for the Cyclops, a giant with one eye in the middle of its forehead. This copepod has one eye, but it's no giant. Most copepods are no more than 1/5 inch (5 mm) long. Like daphnia, they are distantly related to shrimp. Copepods use their legs to filter smaller zooplankton from the water.

Members of the crustacean family that includes lobsters and shrimp, copepods are an important winter food resource in the spring pool.

Two other important spring pool animals are hydra and fairy shrimp. The ancient Greeks told stories about a nine-headed serpent called the hydra. For every head the Greek hero Hercules cut off, the legendary hydra grew two new ones. The pond hydra doesn't have nine heads. It gets its name from the **tentacles** at the end of its tube-shaped body, and from its ability to grow new tentacles when one is lost or damaged. Relatives of the jellyfishes, hydra use their tentacles to stun prey and move it toward the mouths at the center of their tentacles. When threatened, hydra pull in their tentacles and contract into small blobs until danger has passed.

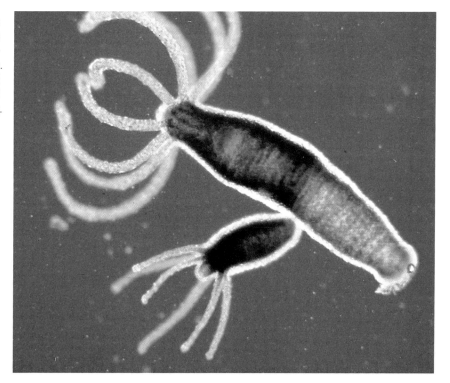

Like their larger relatives, the jellyfish, hydra use stinging tentacles to stun their prey—mostly daphnia and copepods. Here a new hydra splits off from its parent in a kind of reproduction called budding.

With its rippling limbs and neon-spotted tail, the fairy shrimp is one of the spring pool's most beautiful and unusual creatures. It swims upside down, using eleven pairs of fringed legs to fan food toward its mouth. Unlike the daphnia and copepods, which live in fresh water all over the world, some kinds of fairy shrimp are found only in spring pools. Their brief life cycle lasts only as long as the pool. When the waters shrink, fairy shrimp lay their eggs and die. The female's dark blue egg sacs sink to the bottom of the pool, and dry out as the pool's waters evaporate. The next spring, when the pool fills, the dried-up eggs will hatch.

Fairy shrimp use feathery limbs called swimmerets to propel themselves through the water on their backs. Because they are easy to spot and they swim slowly, they have a better chance of survival in a spring pool, away from fish predators.

ON THE BOTTOM

Seldom longer than half an inch, filter-feeding fingernail clams are themselves an important food source for other spring-pool species. They survive dry spells by burrowing into the mud.

The mud and leaves of the pool bottom offer a hiding place and a rich source of food. Larvae of dragonflies and crane flies crawl along the bottom in search of prey. A scoop of the bottom mud yields tiny snails, fingernail clams as small as pinheads, and long, strange-looking worms.

Crawling along the pool bottom is a creature that looks like a mossy piece of bark or a twig or pebble covered with tiny snails. This is the larva of the caddisfly. The larvae build cases to protect themselves. Like a caterpillar weaving a cocoon, a caddisfly larva spins silk into a tube. It then makes a sticky liquid that it uses to glue decorations to its tube. Some species decorate their cases with tiny shells of aquatic snails, moss, or grains of sand. One species uses twigs to construct a case that looks like a log cabin. The resulting burrow is portable, offers camouflage from predators, and anchors the larva in the pool. Adults are small brown, mothlike insects.

You would recognize this adult as a dragonfly, but it spends its early life stages in a less familiar form—as an aquatic nymph (right), feeding on tadpoles, insects, and other larvae.

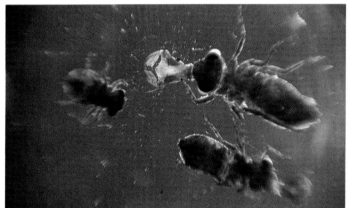

The caddisfly is among the spring-pool inhabitants that demonstrate the pool's importance as a nursery. The caddisfly larva gets protection from predators by spinning a silken tube that it camouflages with twigs, algae, pebbles, or even tiny snail shells.

SUMMER MUD

The temporary nature of the spring pool affects the animals living there. If they don't complete their larval development fast enough, they can be left high and dry—like these tadpoles.

A spring pool isn't the perfect environment. It has unique problems and challenges. During the summer each species must race to produce a new generation. The young must reach adulthood before the pool dries up at summer's end. If they do not mature in time, frogs and salamander larvae might end up stranded in the shrinking mud hole, easy pickings for birds and raccoons. By summer's end, most of the spring pool's inhabitants have flown or crawled away. Many insects have mated, laid their eggs, and died—their short life cycle over.

Some spring pool animals have found ways to beat the summer heat. Spadefoot toads and spotted turtles go into a summer form of hibernation called estivation. In this state of suspended animation, the animal needs less food and spends less energy. It's easier to stay cool when you don't have to move around looking for food or escaping predators. If they don't estivate, frogs and turtles bury themselves in the mud to keep cool.

Snails and crustaceans produce mucus and coat themselves with it to keep from drying out. Small animals called rotifers carry the mucus idea further. They seal themselves inside a tough outer coat called a **cyst** that protects them from very dry, hot, or cold weather. Once warm spring rains arrive, the cyst dissolves, and the rotifer goes about its business of eating zooplankton.

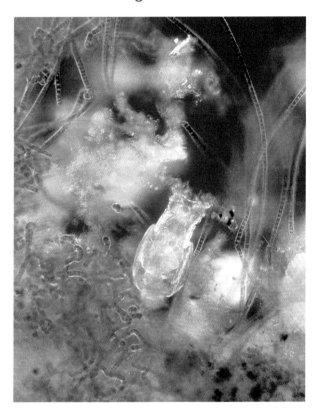

There are about 1,700 kinds of rotifers in the world, but most of us scarcely notice. While magnified here, their small size (.002 to .08 of an inch) makes them nearly invisible to the naked eye, but their importance in the aquatic food web is immense. Tiny beating cilia or hairs surround their heads and bring food into their mouths.

WINTER ICE

As cold weather sets in, animals in spring pools that haven't evaporated face a new set of problems. Food sources dwindle, a lid of ice seals off the pool from fresh oxygen, and there is the chance that the shallow pool could freeze solid.

Like many vernal pool animals, daphnia prepare for winter by laying special eggs. During the summer, females lay unfertilized eggs that hatch into more females. As winter approaches, females start laying eggs that hatch into males. After mating with these new males, females produce egg cases that sink to the pool bottom. These "time capsules" contain tough-skinned eggs that can winter over beneath the ice and hatch when warm weather returns.

Some frogs that winter over on land have developed an unusual adaptation to winter in the spring pool: they can survive being frozen solid! When the tem-

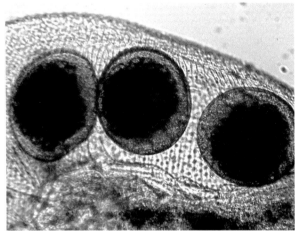

This close-up view shows an egg-case with the daphnia's eggs. Daphnia eggs go into a resting stage under winter ice, or in a dried-up pond, then hatch when spring rains return.

perature of an animal falls below freezing, sharp-edged ice crystals form inside each cell inside the body. When the sharp ice bursts the cell walls, the animal dies. But cells of gray tree frogs and spring peepers contain a special chemical. When ice crystals form in these frogs' cells, the chemical gives the ice crystals smooth edges so that the cells don't burst. A frog that freezes solid in a spring pool can thaw out unharmed.

DANGER SIGNS

Factory and car exhausts contribute to the alarming buildup of acid rain across North America. Sulfur dioxide and nitrogen oxides from the exhausts combine with water to form sulfuric and nitric acids. The resulting acid rain falls into rivers, lakes, and ponds where it may kill wildlife larvae and adults.

While some frogs and salamanders seek out new pools if theirs becomes crowded, most are faithful to the breeding pools where they hatched. Year after year, they travel up to a half mile to reach the pools. Driven by instinct and steered by an internal compass, they cross fields, meadows, and roads. They are determined to reach their old pool, even if the site is now a parking lot or fast-food restaurant. Sadly, the pool was long ago bulldozed and filled with concrete, but the salamanders still make the journey.

Spring pools are disappearing, and one cause is construction. The spread of cities and suburban development have made the spring pool an endangered **habitat**. Pools may suffer too when sulfur from factory smokestacks and au-

Spring-pool sites like these, in New England and other parts of the country, are disappearing as urban and suburban building spreads.

tomobile exhausts combines with water in the clouds to form acid, which falls as rain. This acid quickly builds up in the shallow waters of a spring pool and may prevent many frog and salamander eggs from hatching. If the pond becomes acid enough, the **algae** at the start of the food chain will die. Without algae, the pool will no longer be able to support life.

As spring pools vanish, animals lose their highway through the woods. Spring pools are part of the "wildlife corridor" linking larger areas of the wild that have been cut up by highways and growing cities. These corridors allow animals to move in search of food, shelter, and mates. Without them, wild animals are cut off from wilderness. They must adapt to urban life or move on. If they don't, animals may starve or fail to find mates. Either way, the future for a species that loses its habitat is bleak.

A habitat cannot be endangered without endangering the things that live there. Spring pools are a microcosm—a tiny world. Like our planet, they are fragile. And, like the inhabitants of the spring pool, we humans have nowhere else to go when our habitat—Planet Earth—is spoiled.

Epilogue: Slow—Salamander Crossing!

In the late 1960s, townspeople in Germany and England became alarmed at the number of toads they saw squashed by cars when the animals crossed roads in the spring to mate. People formed "bucket brigades" and carried toads across roads in plastic buckets. "Toad Crossing" signs cropped up beside country roads across Europe.

Then some German bucket-brigaders had the idea of installing tunnels so the toads could cross under roads. A company donated pipes like the ones used to drain water from airport runways. The plastic concrete used in the pipes wouldn't absorb water. Once wet, the tunnels stayed wet—just right for migrating toads.

Toad-tunnels were installed at heavily used toad crossings and were such a success that the idea was brought to the United States. The organizers contacted Jerry Bertrand of the Massachusetts Audubon Society, who told them the town of Amherst annually closed one of its roads for migrating salamanders. The town agreed to install the pipes to try to protect the salamanders.

This salamander tunnel in Amherst, Massachusetts, is part of a salamander success story. The tunnel protects migrating salamanders as they journey from winter hiding places to spring pools to mate.

On a cold March night in 1988, two dozen people turned out to see if the salamanders would use tunnels that had saved European toads. But the night was too cold and few salamanders appeared. On the next rainy night, while over two hundred people huddled to watch, the male salamanders began their trek to the spring pool. They crawled past fences and entered the tunnels that ran beneath a busy street. A few minutes later, they emerged from the other end. The tunnels worked!

By pulling together, people from opposite sides of the globe had found a way to do something extraordinary. Together, they saved some toads and salamanders. Their tunnel idea is spreading—people in towns in Asia are building tunnels to try to save their toads and salamanders, and in some locations the tunnels can work. By looking at the problems in our own towns and working together, perhaps we can also save the spring pools before they disappear.

Keeping a Spring Pool Diary

In any natural habitat, it's important not to disturb the animals more than necessary to get a good look. These observers used a net to capture a spotted salamander, which they then returned to the cover of some leaves.

If you live near a spring pool and can visit it often, you may want to start a diary of the pool to describe the animals and behavior you observe. If you keep a careful record, your observations could even help protect your pool. In some states vernal pools can be "certified"—that is, recorded and registered with the state. Certified pools are then protected by the state's laws, and sometimes by town or local laws, too. Laws about spring pools and other wetlands are different from state to state and town to town. To find out if your state has laws protecting spring pools, and whether there is a program to certify new pools, write to your state's Fish and Wildlife Department.

To keep a diary, record this information each time you visit: date, time, length of visit, depth of water (measure in the same spot each time), temperature of the air and water, and whether or not it rained. You will also want to keep track of any animals, including eggs and larvae, that you see. Check the appearance of nymphs and larvae as they change from week to week. If you are lucky, you may witness the transformation of a nymph into an adult dragonfly.

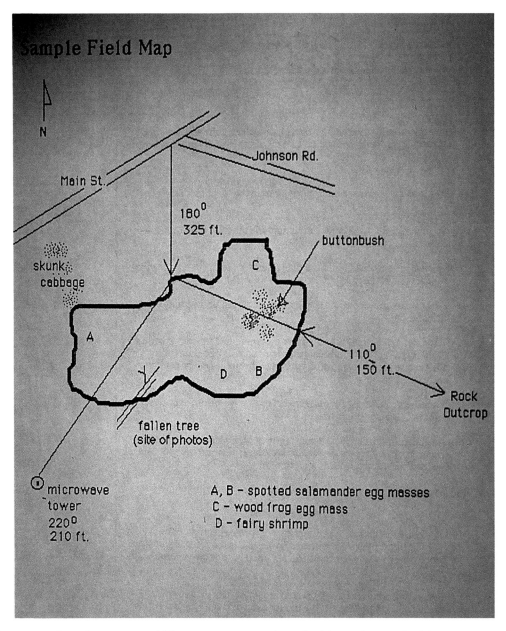

In Massachusetts, and in other states, wildlife organizations such as the Massachusetts Audubon Society offer classes in spring-pool certification. Participants learn how to recognize a spring pool by the signs of the wildlife they find there, and how to survey and draw a field map showing the pool's location. These programs have helped save some spring pools.

IF YOU LIVE IN A CITY

If you don't live near a spring pool, contact the Audubon Society in your area or the education department of your regional Fish and Wildlife Service (listed under "U.S. Government" in the telephone book). Tell them you are interested in spring pool ecology and ask whether they sponsor day hikes in your area.

If you can, bring back a few leaves and a jar of mud from the bottom of a dried-up pool. Take only a small sample (one cup) so that you damage the pool bottom as little as possible. Put the mud in a fish tank or aquarium. Cover the mud with tap water that has been allowed to stand for a day so that the chlorine in it will have evaporated. Put the tank in a warm, but not too hot place. Before long you may see fingernail clams and small freshwater snails sliding up the glass. Add the leaves. After a few minutes you may see tiny blobs unfold into tentacled hydra.

EXPLORING SPRING POOLS ON YOUR OWN

To observe a spring pool, you don't need much more equipment than your eyes, your ears, and plenty of patience. If you plan to study your pool and keep a record of it for school or as a hobby, you might want some of the equipment listed here. Most is inexpensive and can be found in local stores.

- A waterproof flashlight.

- A dip net. You can use a metal kitchen sieve or a net (the finer the mesh, the better) from an aquarium supply store. If you replace the handle with a length of dowel or broomstick, your net will be easier to use.

- A glass jar for examining small animals, such as fairy shrimp.

- A white plastic spoon or medicine dropper for examining tiny animals.

- An enamel or plastic pan, 2 or 3 inches (5 or 7.8 cm) deep, for viewing and photographing pond life up close (a white or light-colored pan provides the best contrast).

- A magnifying glass.

- A thermometer to measure the temperature of the air and water.

- A yard- or meterstick to measure the depth of the pool.

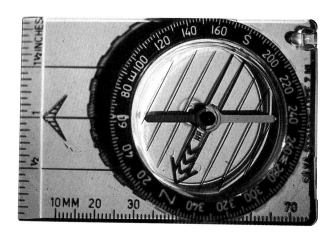

- A compass, if you want to draw a map showing the pool's location.

- A camera. Taking pictures is one of the best ways to keep a record of your visits to the pool. It is worth saving up for or borrowing a 35-millimeter camera, especially if you are interested in developing your skills as a nature photographer.

- A portable tape recorder to preserve the songs of frogs and toads. Try to find one designed to be used outdoors.

- Water-testing kits. Your pet store may carry a kit for testing the level of acid (pH) in water. Biological supply companies carry kits that measure the amount of oxygen dissolved in water. Your science teacher may have the address of a supplier.

BE CONSIDERATE AND BE SAFE

If you want to explore spring pools, follow these guidelines to make sure you don't endanger any animals—or yourself.

1. It is a good idea to go with a friend. If you want to explore a pool in the wild, always go with an adult. Don't go into the woods in hunting season, and make sure you get permission before you enter private property.

2. Carry a map and compass and tell others exactly where you are going and when you plan to be back.

3. Remember, the woods can be very cold in the spring, and if you get wet, it will feel even colder. Dress warmly and wear knee-high waterproof boots. A high-calorie snack such as chocolate or trail-mix will also help keep you fueled and warm.

4. Walk carefully so that you don't trample the edge of the pool or disturb the nests of animals such as turtles.

5. Try not to disturb any egg masses. If you lift them to the water surface to photograph them, make sure you keep them attached to their twig or branch and put them back under water when you have finished so that the eggs don't dry out, become exposed to predators, or sink to the bottom.

6. Handle animals only as much as you need to in order to photograph them. Salamanders and other larger animals can be lifted with a net into an enamel pan filled with water. Fairy shrimp and insects are easier to photograph in a small glass jar.

7. Always keep animals covered with water and, if you are visiting the pool during the day, in the shade.

8. When you have finished photographing or drawing animals, **never** pour them back into the pool! Instead, lower the jar or pan into the water. Let the water fill the jar slowly, then gently tilt it over so the animals can swim out on their own.

9. If you move any plants or rocks, replace them before you leave. Be especially careful to leave egg masses covered with water.

10. Because spring pool habitats are fragile and many of the species that live in them endangered, do not take any animals away from the pool.

FIELD GUIDE
to Spring Pool Ecology

The entries that follow will help you identify some of the animals you might find at a spring pool. Obligate species are marked with an asterisk (*). Scientific names (genus and species) appear in italics after the animal's common name. In a few cases a larger division (family or order) is given because more than one genus may visit a spring pool. Lengths given are from head to tail unless otherwise specified. Descriptions of the animals' diet tell what they feed on in ponds.

Because this book is based on spring pools in eastern Massachusetts, most of the animals described are native to the East Coast of the United States. To find out about salamander, frog, toad, turtle, and invertebrate species where you live, ask your librarian or bookseller to recommend field guides for your region or write to your local Audubon Society for a list of native spring-pool species. This guide lists principal vernal-pool species but not all visitors.

INVERTEBRATES

CADDISFLY

Families Phryganeidae and Limnephildae length: larvae, ½ to 1½ inches (1.3 to 3.8 cm); cases ½ to 2 inches (1.3 to 5 cm)

Telltale features: Tube-shaped, spiral, or log-cabin constructions of plant material, sand, or occasionally small snail shells or other animal matter. Adults are brown or gray mothlike insects. Life cycle: Eggs hatch when pool fills with water in spring or fall. Larvae build elaborate cases and live inside them for several weeks or months. At the end of that time they seal themselves inside, emerging one to several weeks later as adults. Adults emerge in late spring and spend the summer in a dormant state. After mating in late autumn, females deposit eggs on overhanging twigs or plants, or on the bottom of dry pools. Diet: Larvae: caddisfly larvae are one of the main predators of salamander larvae in vernal pools. Adults: short-lived and eat little, if anything.

DAMSELFLY
Families Aeschnidae (darners) and Libellulidae (skimmers)
Genera *Sympetrum*, *Libellula*
length: ¾ to 4¾ inches (1.9 to 12.1 cm)

Telltale features: Large compound eyes nearly cover the head. Four wings extend horizontally to the sides. The long legs are used to capture prey in flight. Life cycle: Females deposit eggs in or around water. When grown, larvae leave the water and transform into adults. Adults and larvae don't resemble each other closely. Diet: Larvae: aquatic predators which eat insects, tadpoles, and sometimes fish. Adults: other insects.

DIVING BEETLE
Family Dytiscidae
length: up to 1¼ inches (3.2 cm)

Telltale features: Females have grooved, olive green backs; males have smooth green backs with white markings and sticky pads on their front legs for holding on to females during mating. Life cycle: Eggs laid inside the stems of water plants hatch into larvae with long, pointed jaws and oarlike legs for swimming. They float to the surface tail-first, and take in air through the tips of their abdomens. After a year, larvae leave the pool, dig burrows in the soil, and remain underground for four weeks while pupating. Diet: Larvae: snails, worms, insects, tadpoles, salamander larvae. Adults: Almost any insect or larva in the pool, including tadpoles.

DRAGONFLY
Families Calopterygidae, Lestidae, Coenagrionidae, Cordulegastridae
length: 1 to 2 inches (2.5 to 5.1 cm)

Telltale features: Large compound eyes which bulge to the side. Four wings extend vertically toward the rear. The long legs are used to capture prey in flight. Life cycle: Females deposit eggs in or around water. When grown, larvae leave the water and transform into adults. Larvae and adults don't resemble each other closely. Diet: Larvae: aquatic predators which eat insects, tadpoles, and sometimes fish. Adults: other insects.

FAIRY SHRIMP*

Order Anostraca, genus *Eubranchipus*
length: ½ to 1½ inches (1.3 to 3.8 cm)

Telltale features: Transparent orange body, two hornlike antennae. Body has eleven pairs of feathery legs and ends in a long forked tail. Life cycle: In many species, eggs must dry and freeze before they can hatch when the pool refills. Once hatched, fairy shrimp go through several larval stages before maturing. Adults mate and eggs are produced which—when mature—fall to the bottom of the vernal pool. Adults die after egg production is complete. Diet: Swimming on its back, the fairy shrimp uses its legs to fan algae, dead matter, and zooplankton toward its mouth.

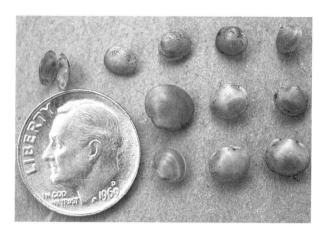

FINGERNAIL CLAM

Genera *Sphaerium*, *Musculium*, or *Pisidium*
length: ⅛ to ½ inch (.3 to 1.3 cm)

Telltale features: Grayish white or tan, ranging in size from the head of a pin to the size of a dime. Life cycle: Adults bear live young, releasing them from a pouch in their gills onto the pool bottom. During

droughts and winter months, adult and juvenile clams burrow into the mud of drying pools until wet conditions return. Diet: Fingernail clams strain the water for a meal of algae and zooplankton.

WATER FLEA

Genus *Daphnia* (more than one species)
length: 1/10 inch (.25 cm)

Telltale features: Rounded, transparent body with pointed "beak" and long,

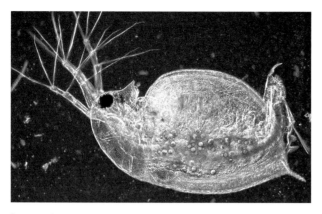

branching antennae used to row through the water. Life cycle: Unfertilized eggs carried in a pouch beneath the shell hatch into females in the summer. During extreme heat and cold females produce a special case containing hardy eggs. The case sinks to the bottom of the pool where the eggs remain dormant until spring. Diet: Uses hairs on its legs to strain bacteria and algae from the water.

WHIRLIGIG BEETLE
Genus *Gyrindae* (more than one species)
length: 1/5 to 2/5 inch (.5 to 1 cm)

Telltale features: Shiny-black, oval beetles spin rapidly on the surface in a spiral pattern. Adults have two pairs of eyes for seeing above and below the water line. Life cycle: Females lay eggs in small batches on the leaves of underwater plants. Larvae are segmented, with long hairs that function as gills. Larvae emerge from the water after July to build cocoons out of mud. Adults emerge in late summer and early fall. Diet: Insects trapped in the pool's surface film.

AMPHIBIANS

BLUE-SPOTTED SALAMANDER*
Ambystoma laterale
length: 4 to 6 inches (10.2 to 15.2 cm)

Telltale features: Uneven blue or pale blue spots along the sides of the body and tail. May have spots on the back. Life cycle: Female lays up to 500 eggs in loose masses of about 30, attached to submerged plants or leaves on the bottom of the pool. Eggs hatch in April and May, and adults are ready to leave the pool in September. Diet: Beetles and other insects, centipedes, and worms.

FOUR-TOED SALAMANDER
Hemidactylium scutatum
length: 2 to 4 inches (5 to 10.2 cm)

Telltale features: Bright white underbelly has contrasting black spots. Only four toes on hind limbs (other species have five).

Body appears pinched at base of tail. Life cycle: Singly or together, females lay their eggs on land in a hummock of moss or grass, often on the edge of a spring pool. Once hatched, larvae enter the pool and transform into juveniles in 6 to 8 weeks. Diet: Small insects, such as springtails and flies, and small spiders.

pool by later summer. Diet: Larvae: crustaceans and insects. Adults: beetles and other insects, centipedes, and worms.

JEFFERSON'S SALAMANDER*
Ambystoma jeffersonianum
length: 5 to 7 inches (12.7 to 17.8 cm)

Telltale features: Dark brown or gray above with flecks of blue on the sides, legs, and tail. Underbelly is lighter. Life cycle: Female lays an amber mass containing 10 to 75 eggs and attaches it to submerged plant stems. Eggs hatch in the spring and adults are ready to leave the

MARBLED SALAMANDER*
Ambystoma opacum
length: 3½ to 5 inches (9 to 12.7 cm)

Telltale features: White or silver dumbbell-shaped spots along the back. Underside is black without markings. Life cycle: The female lays about 150 unattached eggs in dry pools during August and September and guards them until the pool fills and covers them. Once covered, the eggs hatch, and the larvae winter over beneath the ice. Diet: Larvae under the ice are very sluggish, but they do grow, slowly, probably eating copepods. Older larvae eat tadpoles, the larvae of other salamanders, water fleas, and other aquatic invertebrates. Adult diet is similar to that of other salamanders.

SPOTTED SALAMANDER*
Ambystoma maculatum
length: 6 to 8 inches (15.2 to 20 cm)

Telltale features: Body is dark blue or black with a row of yellow spots along each side. Underbelly is lighter in color. Life cycle: The female attaches egg masses to submerged twigs or stems. The matrix can appear clear, cloudy, or green due to the

presence of algae in the eggs. Eggs hatch in April and May, and adults are ready to leave the pool in July and August. Diet: Larvae: crustaceans and insects. Adults: beetles, centipedes, spiders, slugs, and snails.

AMERICAN TOAD
Bufo americanus
length: 2 to 4½ inches (5 to 11.4 cm)

Telltale features: Dark patches on back contain one or two warts each. Chest and belly have dark spots. Life cycle: During April and May, female lays 4,000 to 12,000 black eggs in long spiral strands wrapped around plant stems in shallow water. Eggs hatch in 3 to 12 days. Tadpoles transform

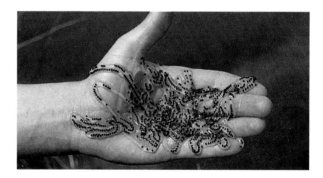

into adults between June and August. Diet: Tadpoles: algae. Adults: many flying insects; large toads also consume small snakes and even mammals, such as mice. Song: Long trill that lasts from 6 to 30 seconds. (Caution: *When threatened, toads puff up and release an irritating secretion through their skin. Although it is not dangerous, you should wash your hands after handling a toad and be sure to avoid touching your eyes and mouth until you do.*)

FOWLER'S TOAD
Bufo woodhousei fowleri
length: 2 to 5 inches (5 to 12.5 cm)

Telltale features: Dark patches on back, each with three or more warts. Body color ranges from tan to dark brown. No spots on underside. Life cycle: During late May and June, female lays 4,000 to 12,000 eggs in long strands around plant stems in shallow water. Eggs hatch in 3 to 12 days. Tadpoles transform into juveniles between August and September. Diet: Same as for American toad. Song: A nasal sound like a goat bleating.

GREEN FROG
Rana clamitans
length: 2½ to 5 inches (6.4 to 12.7 cm)

Telltale features: Two raised ridges down the back; bright yellow spots ringed with black located on each cheek below and just behind the eye. Life cycle: Female lays several thousand eggs in a loose mass attached to submerged stems or spread over the surface of the pool. Eggs hatch between April and June; tadpoles take three months to transform into adults. If conditions are unfavorable, tadpoles can winter over beneath the ice and transform into adults the following spring. Diet: Similar to that of wood frog. Song: A loud twang like a rubber band or banjo string being plucked.

SPRING PEEPER
Pseudacris [Hyla] crucifer
length: ¾ to 1¼ inch (1.9 to 3.2 cm)

Telltale features: An irregular cross-shaped marking on the back. Body ranges from tan to olive. Body thinner and legs longer than

other species. Life cycle: Female lays 800 to 1,000 eggs onto submerged plant stems or onto the bottom of the pool. Eggs hatch in 6 to 12 days in late March or April, and tadpoles transform into adults in July. Diet: Similar to that of wood frog. Song: Loud, fast, high-pitched peeps.

kinds of insects, worms, snails, and slugs. Song: A call like a duck's quack.

REPTILES

WOOD FROG*
Rana sylvatica
length: 1½ to 2¾ inches (3.8 to 7 cm)

Telltale features: A dark brown mask covering both eyes. Life cycle: The wood frog is terrestrial except during the brief breeding season. The female lays about a thousand eggs in a mass attached to submerged stems or on the bottom of the

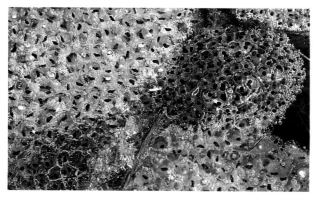

pool. Tadpoles hatch in April and transform into juveniles between June and August. Diet: Tadpoles: algae, bacteria and decaying matter. Adults: many

PAINTED TURTLE
Chrysemys picta picta
length: 3⅞ to 9⅞ inches (9.8 to 24.9 cm)

Telltale features: Patterned red or yellow border on dark green or black shell. No markings on underside. Often found basking and feeding in larger vernal ponds. Life cycle: After mating in late spring, the female digs a bottle-shaped nest in the ground and lays 2 to 20 creamy white eggs that hatch in 9 to 12 weeks. Diet: Algae, aquatic plants, insects, crayfish, snails, slugs, zooplankton.

SNAPPING TURTLE
Chelydra serpentina
length: 8 to 12 inches (20.3 to 30.5 cm)

Telltale features: Large head with hooked beak. Tail is long with sawlike teeth. Life cycle: Mating occurs in the spring or summer. The female lays 20 to 80 eggs in a nest close to the pool. Eggs hatch in late summer to early fall. Young sometimes spend the winter in the nest. Diet: Aquatic plants, worms, tadpoles in the spring pool, ducklings and fish in ponds.

SPOTTED TURTLE
Clemmys guttata
length: 3½ to 4⅞ inches (8.9 to 12.2 cm)

Telltale features: Yellow polka dots on smooth, dark shell. Life cycle: Mating occurs from March to June; the female digs a bottle-shaped nest in the ground and lays 3 to 5 eggs in it. Young hatch in 10 to 12 weeks. Diet: Worms, insects, spiders, slugs, snails, frogs, grass, and algae.

OTHER VISITORS

The species listed below are three of a wide variety of birds and mammals that visit the spring pool. The great blue heron is a top predator at the spring pool. Other birds that may visit the spring pool are ducks, geese, shorebirds, and songbirds. Raccoons can be seen visiting the pool with their young. Some of the other mammals that use the spring pool include mice, muskrats, shrews, voles, and weasels.

GREAT BLUE HERON
Ardea herodias herodias
length: 42 to 50 inches (107 to 127 cm)

Telltale features: Slate blue above, dark below. Long bill and legs. Holds its neck in an S-curve while watching for prey in the shallows. Life cycle: Females lay 3 to 6 blue or blue-green eggs in a messy nest of twigs high in a tree. As many as 150 pairs will nest in a marsh or swamp. Chicks hatch in 4 weeks. Diet: Frogs, salamanders, crayfish. To feed their young, herons migrate to meadows to catch grasshoppers, mice, and snakes.

WOOD DUCK
Aix sponsa
length: 17 to 20 inches (42.5 to 50 cm)

Telltale features: Male has white markings on sides and top of black head; swept-back crest. Back is blue-black; white flecks on brown breast. Sides are tan. Females are chiefly gray and tan. Life cycle: Females lay 10 to 15 tan-colored eggs in a nest in a hollow tree. Hatchlings jump from the nest to the ground and follow their mother to water. Diet: On land, wood ducks feed on seeds, berries, nuts, and acorns, but they visit the spring pool to feed on beetles, mayflies, and tadpoles.

RACCOON
Procyon lotor
length: 28 inches (71 cm)

Telltale features: This well-known mammal is recognized by its black mask and ringed tail. Since raccoons are more active at night, you are more likely to find their tracks, like small handprints, along the edge of the pool. Life cycle: Females bear litters of up to 7 young in the spring. Young are weaned in summer and independent by autumn. Diet: Raccoons will eat almost anything but visit spring pools in search of insects and frogs and their larvae.

Glossary

Abdomen (*AB-duh-mun*)—When describing insects, crustaceans and some other invertebrates, the lower section of the body.

Algae (*AL-jee*)—Free-floating, simple aquatic plants, often single celled, that are an important part of the food chain in fresh- and saltwater habitats.

Amphibian (*am-FIB-ee-un*)—From Greek words meaning "two lives." A vertebrate with a moist skin, usually scaleless, that has a complex life cycle, including a metamorphosis. Frogs, toads, and salamanders are amphibians.

Animal—A form of life separate from bacteria, fungi, and plants; usually animals can move freely, receive and respond to sensations, and eat. The walls of their cells are different from those of plants. It's important not to confuse animals with mammals. Mammals are animals that have warm blood and hair; with some exceptions, they bear live young and nurse them with milk. All mammals are animals, but there are many kinds of animals that are not mammals, for example, sponges, worms, spiders, fish, birds, and snakes.

Aquatic (*uh-KWAH-tic*)—Of or dwelling in water.

Cloaca (*clo-AY-cuh*)—A single opening in the body of all salamanders that leads from different organs to allow the animal to digest food, get rid of wastes, and reproduce. Other amphibians, fishes, reptiles, and all birds and some mammals have cloacae (*clo-AY-cay*).

Crustaceans (*krus-TAY-shuns*)—A large class of invertebrates, made up of marine and freshwater animals including shrimp, crabs, lobsters, and many microscopic species. Like insects, crustaceans have hard, flexible outer shells.

Cyst (*CIST*)—A sac around a tiny organism that is in a dormant state.

Facultative (*FAK-uhl-TAY-tiv*)—Able to live in more than one environment; not dependent on a single habitat (see *obligate*).

Habitat (*HAB-i-tat*)—The area or type of site where a plant or animal normally lives and grows.

Incubate (*IN-kyoo-bate*)—To develop or grow; usually used to refer to the development of fertilized eggs.

Insects A large class of invertebrate animals with bodies divided into a head, thorax, and abdomen; and with three pairs of jointed legs and a hard, flexible outer shell.

Invertebrate (*in-VER-tuh-brate*)—An animal without a backbone.

Larva (*LAR-vah*)—The second stage in the life cycle of an animal that hatches from an egg and undergoes a metamorphosis, especially the wingless, feeding stage of an insect.

Matrix (*MAY-trix*)—The clear or milky jellylike substance that holds an amphibian's egg mass together in the water.

Metamorphose (*met-uh-MORE-fohz*)—To undergo a big change or transformation. The most familiar metamorphosis is that of the butterfly. A butterfly egg hatches into a caterpillar. After growing fat on leaves, the caterpillar spins a cocoon and becomes a pupa that eventually changes into a butterfly.

Metamorphosis, complete (*met-uh-MORE-fah-sis*)—In insects and other animals, development through four distinct stages: egg, larva, pupa and adult.

Metamorphosis, incomplete —In insects, development through three distinct stages: egg, larva and adult, with no pupal stage.

Nymph (*NIMF*)—The young of an insect that undergoes gradual transformation from immature animal to adult (with no pupal stage). Grasshoppers are typical.

Obligate (*AH-blah-gut*)—Dependent on a single habitat for survival; unable to live in other environments (see *facultative*).

Photosynthesis (*foh-toh-SIN-thuh-sis*)—The process by which plants use the energy of the sun to turn water and carbon dioxide gas into simple sugars that are used as food.

Pupa (*PYEW-pa*)—The third life stage (after the larval stage) in the life cycle of insects which undergo complete metamorphosis. After a full internal reorganization, the insect emerges in its adult form.

Species (*SPEE-sheez*)—A group of animals or plants of the same kind. Members of the same species can breed with each other but not with animals of other species.

Spermatophore (*spurm-AT-oh-for*)—A capsule containing sperm.

Tentacles (*TEN-tuh-kuls*)—Long, flexible structures on an animal's head or near its mouth, used for grasping, filtering or stinging.

Terrestrial (*tuh-RES-tree-uhl*)—Of or living on the land.

Thorax (*THOR-ax*)—When describing crustaceans, insects, and some other invertebrates, the section of the body between the head and the abdomen.

Vertebrate (*VER-tuh-brate*)—An animal with a backbone.

Zooplankton (*ZOH-plank-tun*)—Tiny animals, some microscopic, that swim or float in fresh- and saltwater and cannot resist currents or waves.

Bibliography

Amos, William H. *The Life of the Pond.* New York: McGraw Hill, 1967.

Caduto, Michael J. *Pond and Brook: A Guide to Nature Study in Freshwater Environments.* Hanover, N.H.: University Press of New England, 1985, pp. 82–88, 110–111, 197–198.

Colburn, Elizabeth A., ed. *Certified: A Citizen's Step-by-Step Guide to Protecting Vernal Pools,* 4th ed. Lincoln, Mass.: Massachusetts Audubon Society, 1991, pp. 13–24, 47-48, 62.

Ernst, Carl H., and Roger W. Barbour. *Turtles of the United States.* Lexington: University of Kentucky Press, 1972, pp. 71–75, 138–143.

Fellman, Bruce. "A Case of Spotted Fervor." *National Wildlife,* April–May 1990, pp. 12–16.

Klots, Elsie B. *The New Field Book of Freshwater Life*. New York: G.P. Putnam's Sons, 1966, pp. 172–174.

Marshall, Alexandra. *Still Waters*. New York: William Morrow, 1978.

Massachusetts Audubon Society. "Betsy in Wonderland." *Sanctuary*, February/March 1990.

Popham, Edward J. *Some Aspects of Life in Fresh Water*. Cambridge, Mass.: Harvard University Press, 1961.

Roble, Stephen M. "Life in Fleeting Waters." *Massachusetts Wildlife*, Spring 1989, pp. 23-28.

Roth, Charles E. *The Wildlife Observer's Guidebook*. Englewood Cliffs, N.J.: Prentice-Hall, 1982.

Thompson, Gerald, and Jennifer Coldrey. *The Pond*. Cambridge, Mass.: MIT Press, 1984, pp. 94–97, 128–131, 156–157, 161–163, 171, 194–197.

Tyning, Tom. "Amherst's Tunneling Amphibians." *Defenders of Wildlife*, September/October 1989, pp. 20–23.

Read More About It

FOR YOUNGER READERS (ages 7 to 14)

Buck, Margaret Waring. *In Ponds and Streams*. New York: Abingdon Press, 1955.

Cole, Joanna. *A Frog's Body*. New York: William Morrow, 1980.

Hickman, Pamela M. *Bugwise: Thirty Incredible Insect Investigations and Arachnid Activities*. Reading, Mass.: Addison-Wesley, 1990.

Johnson, Sylvia A. *Water Insects*. Minneapolis: Lerner Publications, 1989.

Johnston, Ginny, and Judy Cutchins. *Slippery Babies: Young Frogs, Toads, and Salamanders*. New York: Morrow Junior Books, 1991.

Parker, Steve. *Eyewitness Books: Pond and River*. New York: Alfred A. Knopf, 1988.

Pringle, Laurence P. "Exploring a Pond." In *Discovering Nature Outdoors: A Nature and Science Guide to Investigations of Life in the Fields, Forests, and Ponds*. Garden City, N.Y.: Natural History Press, 1969, pp. 88–110.

Schwartz, George I. *Life in a Drop of Water*. Garden City, N.Y.: Natural History Press, 1970.

FOR OLDER READERS (ages 14 and up)

Marshall, Alexandra. *Still Waters*. New York: William Morrow, 1978.

Martin, Glen. "Spring Fever." *Discover*, March 1990, pp. 71–74. California spring pools.

Morgan, Ann Haven. *Field Book of Ponds and Streams: An Introduction to the Life of Fresh Water*. New York: G.P. Putnam's Sons, 1930.

Roth, Charles E. *The Wildlife Observer's Guidebook.* Englewood Cliffs, N.J.: Prentice-Hall, 1982.

Russell, Franklin. *Watchers at the Pond.* New York: Alfred A. Knopf, 1961.

Tyning, Tom F. *A Guide to Amphibians and Reptiles.* Boston: Little, Brown, 1990.

Index

Acid rain, 31
Aeschnidae family, 43
Aix sponsa, 51
Algae, 21, 31
Ambystoma jeffersonianum, 46
Ambystoma laterale, 45
Ambystoma maculatum, 46
Ambystoma opacum, 46
American toads, 47
Amherst, Massachusetts, 32
Amphibians, 7, 45–49
Anostraca order, 44
Aquatic life. *See* Spring pools, list of inhabitants
Ardea herodias herodias, 51
Audubon Society, 36, 42

Backswimmers, 19
Bertrand, Jerry, 32
Blue-spotted salamanders, 45
Bottom life of spring pools, 24
Bufo americanus, 47
Bufo woodhousei fowleri, 48

Caddisflies, 24, 42
Calopterygidae family, 43
Caterpillars, 7
Certified pools, 34
Chelydra serpentina, 50

Chrysemys picta picta, 49
Clemmys guttata, 50
Cloacae, 13
Coenagrioniadae family, 43
Congressing, 11
Copepods, 21
Cordulegastridae family, 43
Crane flies, 24
Crustaceans, 27
Cyclops, 21
Cyst, 27

Damselflies, 43
Daphnia, 21, 27
Daphnia, genus, 44
Darners, 43
Dependent species, 7
Diary of a spring pool, 34–35
Disappearance of spring pools, 29–31
Diving beetles, 19, 43
Dragonflies, 7, 24, 35, 43–44
Ducks, 50, 51
Dytiscidae family, 43

Ecological threats to spring pools, 29–31
Egg masses, 13, 40, 41
Eggs, 7, 13, 21, 23, 27, 35
Endangered habitats, 29–31
Estivation, 26
Eubranchipus, genus, 44
Exploring spring pools, 37–41

Facultative species, 9
Fairy shrimp, 22–23, 41, 44
Field guide to spring pool species, 42–51
Fingernail clams, 24, 36, 44
Fish and Wildlife Service, 34, 36
Four-toed salamander, 45–46
Fowler's toads, 48
Freshwater snails, 36
Frogs, 15–17, 26, 27–28, 29, 31, 48, 49

Geese, 50
Genes, 13
German attempts to save toads, 32
Gray tree frogs, 28
Great blue herons, 50–51
Green frogs, 48
Gyrindae, genus, 45

Hercules, 22
Hermidactylium scutatum, 45
Herons, 9, 50–51
Hibernation, 26
Hydra, 22, 36

Inhabitants of spring pools, listing of, 42–50
Insects, 7
Invertebrates, 42–45

Jefferson's salamander, 45

Larvae, 7, 24, 26, 35
Lestidae family, 43
Libellula, genus, 43
Libellulidae family, 43
Limnephildae family, 42

Marbled salamanders, 46–47
Massachusetts Audubon Society, 32
Matrix, 13
Mice, 50
Musculium, genus, 44
Muskrats, 50

Nonobligate species, 9
Nymphs, 7, 35

Obligate species, 7, 42

Painted turtles, 49
Photographing spring pools, 41
Phryganeidae family, 42

Pisidium, genus, 44
Pond skaters, 19
Predators, 7, 14, 21
Procyon lotor, 51
Pseudacris crucifer, 48

Raccoons, 9, 50, 51
Rana clamitans, 48
Rana sylvatica, 49
Reptiles, 49–50
Rotifers, 27

Safety rules for exploring spring pools, 40–41
Salamanders, 11–14, 21, 29, 31, 32–33, 41
Saving toads and salamanders, 32–33
Shorebirds, 50
Shrews, 50
Skimmers, 43
Snails, 22–23, 27, 36
Snapping turtles, 50
Songbirds, 50
Spadefoot toads, 26
Sperm, 13
Spermatophore, 11–13
Sphaerium, genus, 44
Spotted salamanders, 11–14, 46
Spotted turtles, 26, 50
Spring peepers, 28, 48
Spring pools:
    bottom life, 24
    coming to life, 5–9
    ecological threats, 29–31
    exploring, 37–38
    frogs and toads, 15–17
    keeping a diary, 34–36
    list of inhabitants, 42–50
    list of visitors, 50–51
    safety rules, 40–41
    salamanders, 11–14
    saving toads and salamanders, 32–33
    summer problems, 26–27

surface life, 19
  underwater life, 21–23
  winter problems, 27–28
Springtails, 19
Summer problems of spring pools, 26–27
Surface life of spring pools, 19
Sympetrum, genus, 43

Tadpoles, 7, 21
Tentacles, 22
"Toad Crossing" signs, 32
Toads, 15, 26, 32–33, 47, 48
Toad-tunnels, 32–33
Turtles, 26, 48–50

Underwater life of spring pools, 21–23

Vernal pools. *See* Spring pools
Vertebrates, 21
Visitors to spring pools, 50–51
Voles, 50

Water fleas, 21, 44
Weasels, 50
Whirligig beetles, 19, 45
Winter problems of spring pools, 27–28
Wood ducks, 51
Wood frogs, 49
Worms, 24

Zooplankton, 21, 27

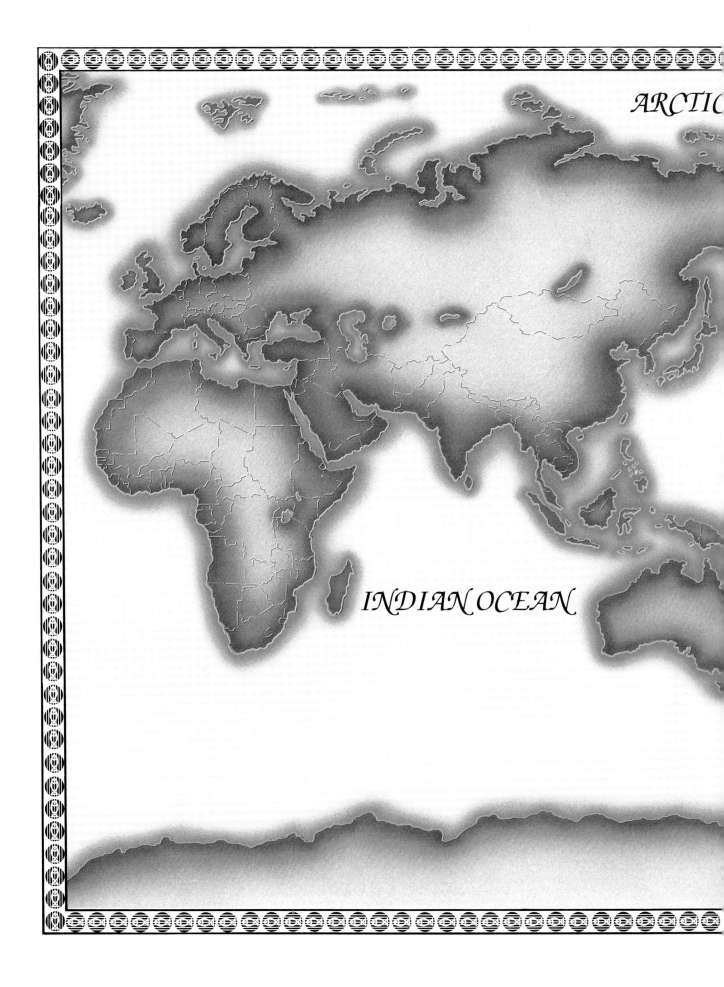

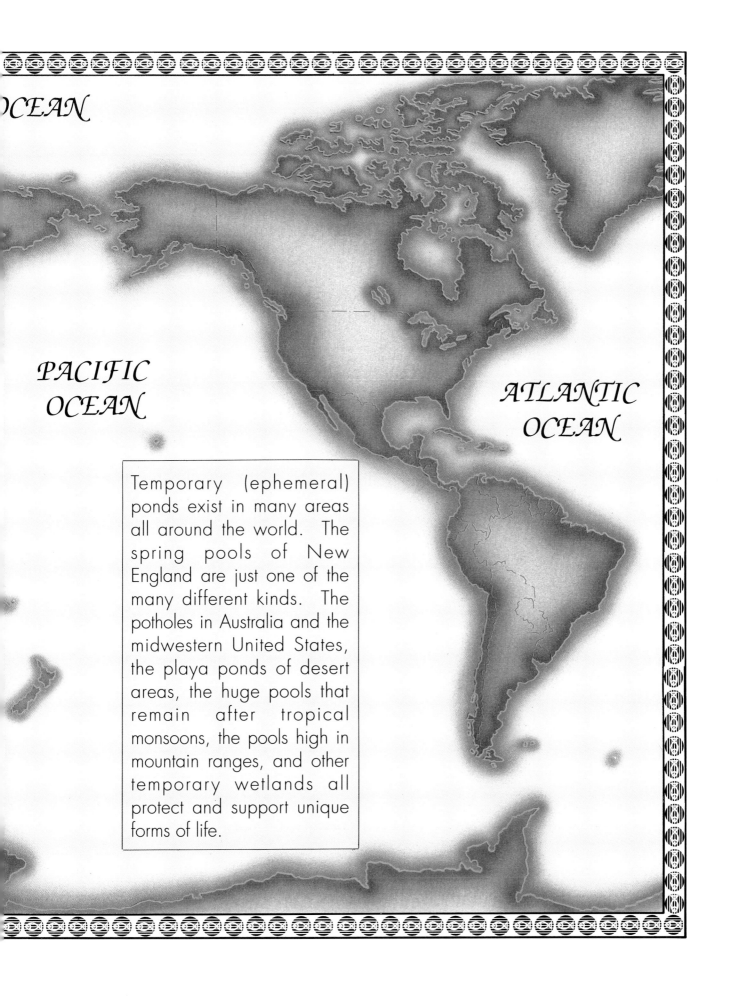

Temporary (ephemeral) ponds exist in many areas all around the world. The spring pools of New England are just one of the many different kinds. The potholes in Australia and the midwestern United States, the playa ponds of desert areas, the huge pools that remain after tropical monsoons, the pools high in mountain ranges, and other temporary wetlands all protect and support unique forms of life.